THE ROUNDEST NUMBER IN THE WORLD

Cacildo Marques

Copyright © 2017 Cacildo Marques
All rights reserved.

ISBN: **978-1979710428**

Marques, Cacildo
The roundest number in the world/ Cacildo Marques.
EpistemeEd , 2017.

40p.
        ISBN: 978-1979710428

I. Arithmetic Operations - Zero. I. Titles   CDD 513

CONTENTS

0	**Alert**	vi
1	**Origins**	1
2	**Error**	6
3	**Bases**	7
4	**Positional**	16
5	**Excursion**	18
6	**Adoption**	21
7	**Impossible**	25
8	**Demonstration**	28
9	**Finishing**	30

Alert

The number zero, the simplest of all, is what scares the most students in elementary school. This happens since the student is informed that he cannot divide any other value for it and also that when multiplied by any amount, however large, it gives result zero.

One of the most necessary tasks of the elementary school teacher is to demystify zero, to frighten away the fear and strangeness that this number causes in children and adolescents. There is no exaggeration in saying that, when the student becomes confident of working with zero without fear, the rest of the Arithmetic becomes easy, as it really should be.

Of course, learning Math requires dedication and completion of steps. Those who are lazy to study and those who want to skip topics will have many more reasons to complain about the alleged difficulty. One confuses, therefore, difficulty with gap.

Enjoy zero, so that your test score is never a zero.

The author.

THE ROUNDEST NUMBER IN THE WORLD

Cacildo Marques

THE ROUNDEST NUMBER IN THE WORLD

1. Origins

Marino's teacher, Olga, of Mathematics, spoke in class about the origin of the writing of numerals.

- One of the hypotheses for the shape of the Hindu-Arabic numerals is the quantity of angles. Does anyone among you know this?

- Hypothesis of the quantity of angles? One of the students asked. - I never heard about it.

- Neither do I, Marino said.

- Yes, the teacher said. - Some researchers suggest that the digits began to be written in a way that obeyed the number of sharp or straight angles they represented. The number 1 was written at an angle; the number 2, with two angles; the number 3, with three angles, and so on.

- How is this, teacher? Show us - student Evander asked.

Olga drew on the blackboard:

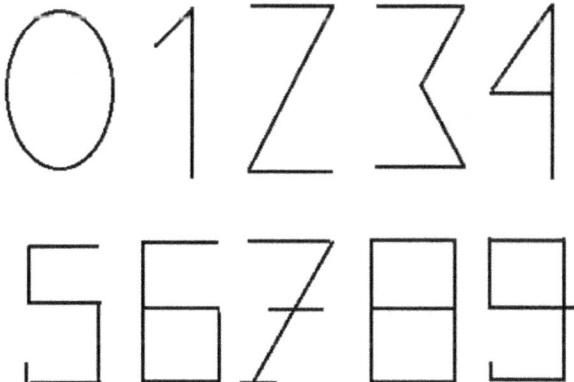

- Hey, teacher, this "seven" got "forced", huh? - the student Ireneaus joked.

- Why, Irenaeus? - Olga asked.

- Because no one writes "seven" with a foot, like that - the student said.

- As you see, not everything is as perfect as you want - Olga said, - but how do you know if seven was not even written like this in its early days? The writing of numbers has undergone many modifications over the centuries.

- Is this explanation of the origin of numbers a serious thing, teacher? - Marino asked.

- Someone saw this coincidence and started to explore, but there is no historical document showing the writing from segments and angles. On the contrary. I'll show you their form on the oldest record.

Olga drew again on the blackboard the ancient Hindu numerals:

- Teacher - Marino said, - one of the figures is missing there, is not it?

- That's right - Olga continued. - You know what it is, do not you?

- I know - Marino said.

- What's it, Evander? - Olga asked.

- Is it ten, teacher? - Evander ventured.

- Are you crazy, Evander? - Olga said, startled. - Where have you seen digit "ten" in Hindu-Arabic numerals? I'm going to give you a problem and I want you to do it individually. Then you will find out what that number is.

Olga wrote on the board:

DIVIDE 267 BY 25.
FIND THE INTEGER QUOCIENT AND THE REMAINDER.

After a few minutes a student raised his finger.

- Teacher, I've figured it out - Leander said.

- How much? - Olga asked.

- Quotient 1 and remainder 17.

- Now take the real proof – the teacher said.

Leander multiplied 1 by 25 and added to the result the remainder 17, getting 42, instead of the amount he needed to get, which would be the dividend 267.

- Oh teacher; I was wrong – Leander confessed.

- Whoever has obtained the result didn't say it yet - Olga said. - I want see everyone to try and, if possible, to find the answer.

- After a few more minutes, Olga ordered that those who obtained the correct result, and had confirmed through the proof, raised their hands. Of the 32 students in the class, 22 raised their hands, and Olga found the result not very encouraging, complaining that in a sixth grade class, although at the beginning of the school year, there were so many students still unable to complete the integer division of 267 by 25.

Olga then did the operation on the chalkboard:

```
267 | 25
-25   10
 017
```

- Now everyone knows the missing number - she said. - What's that number?

- Zero - the students replied in unison.

- You realized that if the second figure in the quotient was another, everyone would have made the division. Why does the error occur? It's because of the little experience of the mankind with the number zero. The symbol of this number, which is written without any angle, is very new, compared to the other figures. It appeared for the first time in a text written in the year 876, while the other nine digit symbols are more than two thousand years old. A few time ago it was discovered in India a document called Bakhshali Manuscript, which bears the earliest record of the figures, from the second century BC. What is curious is that after 9 a little black ball comes, which many imagine being a representation of zero, but it is unlikely to be, being more worth the guess that it is a kind of end point.

- Wow! Evander admired. - The teacher knows dates by heart.

- Important dates like this, I know them - Olga said.

Why is it so important, teacher? - Leander asked.

- Well — Olga said, - because zero is the most important number of all.

- But it's nothing! - Evander wondered.

Olga drew in the chalk the shape of the numbers in the Bakhshali Manuscript.

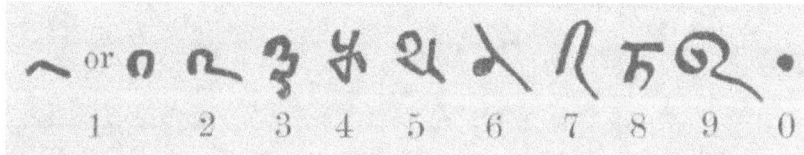

- You make a mistake - Olga said. - It is not true that zero is nothing: what it is, in reality, is the representation of the number of elements of nothing. It is the amount of elements, the cardinality, of the set that has no elements, of the empty set. If I present to you a set without elements and ask "what is this?" you will not answer "zero", but rather, "the void", or "the empty set". But if I ask "how many elements does this set have?" what should you answer?

- Zero! - The class replied.

- That's right - Olga confirmed. - Then you certainly realized that nothingness, or emptiness, is not the same thing as zero. To consolidate this discovery of you, we will divide the class into five teams, and each of them will research and bring in next week something about a facet of that number.

2. **Error**

The themes that the teacher distributed to the six teams were "The history of zero", "Zero in the addition, subtraction and multiplication", "Zero in the division", "Zero in the exponentiation and rooting" and "Zero in the positional system". Each work should be illustrated with a figure containing the shape of zero.

Marino, arriving at the house, soon went to talk to his father, Professor Lalo, about the subject of the last Mathematics class and about the work that his team would have to deliver. Lalo took out a calculator and said, "I'll show you something".

Then he entered zero divided by 50 and presented to Marino the result: zero. Marino said, "Oh, Father. Who does not know that?" Lalo continued: "Wait. See now". At that point he typed 50 divided by zero and the result on the display was "E".

- Hey! What does this "E" mean? - Marino asked, with surprise.

- Do you have no idea? - Lalo asked socratically.

- I think I already know: ERROR. "E" is short for "Error".

- Obviously! - Lalo confirmed. - Its program comes in English, therefore, it is really "error", "E" from "error".

- Dad, when are you going to buy me a calculator?

- For what? What are you going to do with a calculator?

- For making accounts, aack!

- For you to do accounts or to be lazy and forget about mental operations? You're letting me down.

- Sorry, Dad. I'm not going to ask you for a calculator.

- It's good. Unless you mind, an adult is no longer at risk for certain mental defects, such as a child or teenager.

- My teacher really does not want students to use a calculator.

- With reason. But when you're in high school I'm going to buy you a computer.

- I will not have to use yours, Dad?

- No way! You will have your own one and will put into it as many viruses as you want.

3. Bases

Marino would be meeting with his team at the school library soon enough, to look up information on "Zero in the division" in several books. But he obtained from his father to take to his colleagues a few sentences about zero that, according to Lalo, could not be lacking at the work. They were as follows:

(1) - "Zero is a multiple of any number".

(2) - "Zero is no divisor of any number".

(3) - "You should not write a fraction with denominator zero and numerator other than zero, because that fraction does not exist".

In the school library, Marino and his team asked for several books and began to investigate all that referred to divisions involving the number zero. They classified the cases found.

The first situation, it is somewhat common, is the one in which the remainder gives zero. This is the case of the exact division, and Oswald, a member of the team, who used to be very quiet in class, but who was now loquacious, he was who drew the attention for that. They collected two examples of division with remainder zero:

$$
\begin{array}{ll}
\text{A) } 2954 \mid \underline{7} & \text{B) } 31590 \mid \underline{135} \\
15 422 & 459 234 \\
14 & 540 \\
0 & 000
\end{array}
$$

Claudia suggested as the next situation the case of divisions with dividend zero. Oswald noted that this would be the illustration of the phrase "zero is multiple of any number". They wrote the following examples:

$$
\text{A) } 0 \mid \underline{2} \quad \text{B) } 0 \mid \underline{153}
$$
$$
0 0 0 0
$$

In that moment, Claudia drew attention to the fact that when the dividend is zero, only the divisor is not.

Marino was the one who proposed the third situation: the case where the quotient is zero, and the other elements do not. The group was surprised by the idea and almost rejected it. But Oswald remembered that they were producing a draft and could calmly check later whether the idea was correct or not. They could ask Professor Lalo, Marino's father. They did two examples:

A) 7 | 9   B) 5 | 250
    7   0   5   0

But Marino wanted to resolve the issue at that exact moment. He managed to find in the library a book that dealt with "change of number base", an operation involving the entire division by the number of the numbering base to be adopted in the transformation, and in that book he showed his colleagues an example of division in which the quotient showed zero with the other three non-zero elements. It is like when you do division into decimals, - Marino explained, - you write "zero, dot" in the quotient and then add zero in the dividend, if it's smaller than the divisor. Since the division is integer here, it has no "zero, dot", but only "zero". And then we have to write the remainder, which is equal to the dividend. If we make the division by the long process, he said, in the case of 7 divided by 9 we multiply 0 by 9, which yields 0, and we write this 0 under the dividend 7. By doing 7 minus 0, we obtain the remainder 7.

All the teammates were convinced after this explanation.

Claudia proposed that they include in the work this transformation of number base, since it uses the division and involves the number zero.

In the searched book there was the following example:

"Write in base 2, or binary base, the number 73".

And the solution:

```
73 | 2    36 | 2    18 | 2    9 | 2    4 | 2    2 | 2    1 | 2
13  36    16 18    0  9    1 4    0 4    0 1    1 0
 1         0
```
↑ Unit for the base 2

Result: 73 = 1001001(2).

The division in which only the quotient is zero is the last division in the series. One should note that each quotient is divided by 2 in the next operation, until this quotient is zero. The remainders are then taken backwards to form the number on the binary basis, which is a numbering system composed only by the digits 0 and 1.

Adding 1 + 1 in this base corresponds to adding 9 + 1 in base ten, since the first base power is reached after the number 1. In base ten, this power is 9 + 1 = 10; in base two, it is 1 + 1 = 10(2).

The book presented an example for constructing binary numbers from the sum:

```
 0    1   10   11   100
+1   +1   +1   +1   +1 ...
 1   10   11  100   101
```

The last number in the above sequence, (101(2)), is equal to 5.

The book also showed how to transform the binary number into its decimal equivalent.

In the case of 1001001(2), we had:

$1001001_{(2)} = 1*2^6+0*2^5+0*2^4+1*2^3+0*2^2+0*2^1+1*2^0 =$
$= 2^6 + 2^3 + 2^0 = 64+8+1 = 73$

The number $101_{(2)}$ is:

$101_{(2)} = 1*2^2 + 0*2^1 + 1*2^0 = 2^2 + 2^0 = 4 + 1 = 5.$

And the number $100_{(2)}$:

$100_{(2)} = 1*2^2 + 0*2^1 + 0*2^0 = 2^2 = 4.$

Oswald wanted at that moment to deal with the division of decimals, the division with floating point, but Marino recalled that there was still a special situation to be treated, still in the integer division. It was the division in which zeros appear in the quotient interspersed with other figures. They also included the example used by the teacher.

The examples were then:

```
A) 1428 | 7      B) 267 | 25
   028  204         017   10
     0              17
```

It was explained in the paper that, when dividing 14 by 7, in the example (A), obtaining quotient 2 and remainder 0, while lowering the number 2 is noticed that it results in quotient 0 when divided by 7. One writes this 0 in the quotient and then the digit 8 is lowered to continue the operation.

The team's next step was to study the decimal division involving zero.

The first point to note was that, since the dividend and the divisor are either integer or rational, such a division can result in two cases: an exact decimal or a periodic decimal.

They wrote the following examples:

A) 1 |_5_ .... 10 |_5_   B) 12 |_5_   C) 7 |_3_
    0.   0 0.2   20 2.4   10 2.33 ...
                             0   10
                                               1

D) 2 |_7_ ..... 20 |_7_   E) 3 |_40_ ... 300 |_40_
    0.   60 0.2871428...   0.   200 0.075
             40   00
             50
             10
             20
             60

F) 0.7 |_0.08_ .... 70 |_8_
                 60 8.75
                 40
                  0

At that time, the team decided to write the closing of the work, not without stressing that a division can be represented as a fraction and therefore we do not have fraction with denominator zero, since the denominator represents the divisor.

At the exit of the library, and still in the precinct of school, they passed by the little lake where the housekeeper raised ducks and swans. They paused for a moment to watch these birds, when Claudia noticed that a duck was in a nest, laying an egg.

- Look - Claudia said. - The duck is laying an egg.

- I have not seen so far a duck laying an egg, - Julius, another member of group, said.

- But did you see a chicken, right? - Marino asked.

- Chicken, yes - Julius said.

They left, but the image of the egg in the nest of the duck did not leave the mind of Julius, nor of Claudia.

The next day, before entering the class, Marino was called by a classmate to take a look at the sketch of the paper "Zero in Potentiation and Rooting". He noted that this team dealt with only the integers and then suggested them to include an item about powers and roots with decimal numerals. Professor Lalo had already explained to Marino this writing of decimal numerals called "scientific notation", or "notation in powers of ten".

He presented as examples the writings of the numbers 5100000 and 0,000000983. In scientific notation, Marino explained, the dot is placed in the first decimal place, or decimal order, that is non-zero. In these two given numbers the dot goes to the figures 5 and 9, respectively. But then it is necessary to multiply a power of ten that compensates for this change. In the first number the dot jumped six places, going from right to left, from the last place, thus diminishing the value originally expressed. In the second number, the dot jumped seven places, but going from left to right, thus increasing the value of the number. The first number must be multiplied by a power of high value, 10^6, while the second number must receive a multiplier of value much smaller than unity and which will be the power 10^{-7}, with exponents 6 and -7 being the quantity of places jumped by the dot in each number. The result looks like this:

A) $5,100,000 = 5.1*10^6$.
B) $0.000000983 = 9.83*10^{-7}$.

Marino then wrote other numbers for his colleague to turn to scientific notation. They were: a) 18200, b) 2500000, c) 0.00045, d) 0.00789 and e) 0.091.

Also, the roots of numbers involving zeros could be calculated in a more practical way using scientific notation. Marino gave as an example the square root of 90,000:

$$\sqrt{(90{,}000)} = \sqrt{(9*10^4)} = \sqrt{9}*\sqrt{(10^4)} = 3*10^2 = 300$$

and also:

$$\sqrt{(0.00000025)} = \sqrt{(25*10^{-8})} = 5*10^{-4} = 0.0005.$$

The team had already included in the work the question of numbers raised to zero and the power of base zero. They have explained that zero raised to any non-zero number equals zero

$$(0^x=0, \text{ for } x \neq 0)$$

And also that any number raised to zero, including base zero, is equal to 1:

$$x^0 = 1, \text{ for any } x.$$

For many people, the class wrote, it is difficult to accept the fact that zero raised to zero gives 1, this being another case of the strangeness that zero brings to most humans. But there are sure proofs that the fact is consistent.

4. Positional

Colleagues from other teams also approached to show the outline of their work to the Marino's group. Evander's team had written about "Zero in the positional system" and all that was important in this area had been dealt with in the writings of that group. In the beginning they dealt with the nomenclature relative to the positions of the symbols in the writing of numbers: absolute value and relative value of the digit; orders of unity, tens and hundreds; classes of unity, thousands, million, billion, and so on; value of zero to the left and value of zero to the right.

With respect to the idea of positional value, or relative value of the digit, the team wrote that zero has a special role of "keeping position", being the only one who can do that. In a number whose order of magnitude is tens of thousands, if there is no positive quantity for the place of the simple hundred, for example, who will occupy such a position will be zero. Write "three thousands, two hundred and five units" not like

$$3\ 2\ \_\ 5,$$

but how

$$3\ 2\ 0\ 5.$$

The work explained that a zero, written to the right of any number, creates a new number ten times larger. The number 325 after receiving a right zero becomes 3250, number ten times greater than 325. Thus, adding two zeros is multiplying by 100, three zeros is multiplying by 1000, and so on. It also clarified that adding zeros to the left of an integer in no way changes the number. Thus, $25 = 025 = 0025 = ...$

For numbers in decimal notation, with floating point, the situation is different. Adding zeros to the right, in a number with a dot, does not change the value:

$$0.03 = 0.030 = 0.0300 = ...$$

But a zero added to the left, provided that after the dot, divides the value of the old number by ten. Thus, 0.5 is ten times greater than 0.05 and is one hundred times greater than 0.005. The number 0.08 is a thousand times greater than 0.00008.

There was also a lack of information on the fact that zero negative and zero positive are equal, which makes zero a non-negative and at the same time a non-positive number, being the only number with this characteristic.

As if in jest, Marino told Evander:

- For those who doubted that zero is the most important number, the work of your team does not leave zero in bad situation.

- Ah, Marino - Evander said, - work is work; and moreover it was the whole team who prepared the research. I still think zero is not the most important number in the world.

5. Excursion

Teacher Olga and the Science teacher, Andrea, had decided to organize a class excursion to the Forestry Garden. As the tour took place prior to the delivery of the work on zero, Marino and his team decided to note everything they saw on this trip that had some direct relation to the symbol or the idea of zero.

Claudia was the first to make a note of this type, because as soon as the bus, on the way out to the tour, circled the traffic roundabout in front of the school building, she noticed that the roundabout was zero-shaped, "a much big and very round zero", she said.

During the journey they invented an unprecedented game: it was a game in which the participants had to see first-hand car plates containing the digit zero. At the end of the tour, whoever had first seen the most number of plates with this condition would win the game.

Arriving at the Forestry Garden, he who was winning the game was Evander, who until then had sighted the largest number of plates containing the digit zero.

Walking around the flowerbeds, the boys discussed football, and the girls, volleyball. Teacher Andrea joked:

- Wow, you finally stopped talking about zero!

The crowd laughed and continued the conversation. In a little, they reversed the subjects: boys were discussing volleyball, and girls, football.

A small group, however, was talking about cartoon, and yet another was discussing dinosaurs. Andrea approached the latter group, proposing to participate in the conversation.

At lunchtime the whole class sat on the floor in a circle. It was a picnic. But the memory of the work of Mathematics was inevitably:

- We're inside a zero - Evander said.

- Let me take note of that - Claudia amended.

- I hope - Olga said, - that you associate the idea of the symbol zero with that of the circle of friendship.

- And in a time of fellowship, how this lunch is - Andrea said.

- And I'm going to eat a zero - Oswald joked, - showing a manioc flour biscuit.

- Good thing it's a fairly light zero, Marino said.

- Let me note that, too - Claudia added.

At that moment Oswald began to tell a joke. Some other jokes came, until lunch was over.

Some of the students raised the idea that the normal classes of school should be taught in the open air, in places like the one in which they were at the moment, Andrea promised that from then on many classes would be taken outdoors, but that not all subjects could be dealt with the classroom outside in the absence of blackboard and desks.

Andrea also said that the last time she ministered a class in the open air a big uproar happened: a girl sat on an anthill and when she noticed several ants were already stinging.

- Good thing it was an anthill, not a scorpion's nest -

Oswald teased.

- Oh, Oswald, how sadistic you are! - Andrea complained.

- But this was a great practical class in Biological Sciences - Olga said.

- Do not think - Andrea said, - that I programmed that sitting on the anthill. That class would not even be about animals, but about vegetables.

Finally, it was time to end the walk to the Forestry Garden and everyone got on the bus, ready to go home. On the way back they sang a lot and played with the pedestrians in the street, sometimes even against the guidance of the two teachers.

6. Adoption

The Monday of the following week was the day of submission of the work about zero and each team should make a brief statement of the content of what was searched. The first class in the group was Mathematics, and Olga, shortly after the call, was collecting one by one the works elaborated by the teams.

In the next stage, Olga was calling to the front of the classroom a representative of each group, who would have the task of making a slight presentation of the work. Each student had to describe, at the work, by Olga's demands, how his participation in the team was.

First the student Irenaeus, of the group that wrote about the history of zero, came to the front. He said that already in Rome in the sixth century a philosopher named Boethius tried to replace the use of Roman numerals in the schools by the

Arabic numerals, but that did not succeed.

Boethius could not persuade his contemporaries to exchange Roman numerals for Arabic, and the motive must have been that they, while offering some advantage in writing, still did not carry a symbol for zero, which would come in the nineteenth century to be the largest attraction of that system.

Boethius was condemned to death for political reasons and Europe continued to use Roman numerals. It was only in the eleventh century that a French priest named Gerbert de Aurillac began practicing the use of Hindu-Arabic numerals in the European education. Gerbert later became pope, under the name of Sylvester II, and his former followers came to have a greater reason to propagate his ideas. Thanks to Gerbert, Europe was not so far behind in the use of zero, for only a few decades ago this symbol had been identified in some text: the first impression of the number zero is from a Hindu writing of the the prior century. But it was with the Italian Leonardo of Pisa, nicknamed Fibonacci, in the thirteenth century, that the Hindu-Arabic numerals consolidated their success in Europe. Fibonacci, who had this nickname for being the son of the merchant Bonaccio, was the greatest European mathematician of the Middle Age, having been initiated in this science by Arab merchants friends of his father.

With the Arabs, Fibonacci learned to use, among other resources, the fraction trait, a notation that also spread through Europe.

It is easy to see that if Fibonacci had not assimilated the use of Hindu-Arabic numerals, zero included, and had spent his life working with Roman numerals, he would have passed far from the possibility of becoming the greatest European mathematician of his time. As one knows, Roman numerals do not allow for much advancement in calculus.

Since the greatest of the mathematicians of the continent adopted the numbering of the Arabs, there was no choice left for the smaller mathematicians of Europe that follow that choice. So no one else who had common sense tried to resist this adoption.

The next most important step in the history of zero was the invention of the decimal notation by the Scottish mathematician John Napier. He proposed using the comma or dot to separate the integer part of the decimal part into the writing of a number. The British got the point, while continental Europeans preferred the comma. Before this idea what was in use was a complicated system of numbering, with figures within a circle, the decimal orders of a number. For example,

312 ① 4 ② 3 ③ 7 ④ 2

meant

312.4372.

Someone had to come up to simplify this, even considering that this notation of circles already meant a tremendous advance: the great thinkers of Classical Greece did not even conceive of the idea of fraction, much less numerals with decimal orders. Number for the ancient Greeks was positive integer, and end. That is why the letters of the alphabet were convenient to represent them. This history of fractions, decimals, and radicals was not in the minds of even the most imaginative researchers. At that point, a question from one of the students in the class disconcerted Irenaeus.

- Who was the inventor of zero?

As Irenaeus hesitated, with no way out of the question, teacher Olga intervened:

- There is no record of who invented this symbol. The person who threw the symbol to zero may have also wanted to be anonymous, like the inventors of the other digits, which are much older.

- It's like the inventors of the alphabet, is not it, teacher? - Claudia asked.

- The alphabet, the wheel, the ax, the slipper - Olga replied. - But the irony is that the inventor of the alphabet created the instrument so that future inventors could register the authorship of their works. And he, he exactly, was an anonymous.

- He invented the alphabet, but not the copyright - Marino said.

- True - Olga agreed, - but how do you know this issue of copyright?

- My father. He's the one who lives fighting for copyright.

7. Impossible

The next group to have the work presented at the front of the class was the one that studied the theme "Zero in the positional system". The student representing the team began by exposing the nomenclature of the decimal orders (unit, ten and hundred) and the numbering system classes (units, thousands, million, billion, trillion, etc.) to the board.

Then the student explained that in each place, or decimal order, the figure had an absolute value, from zero to 9, and a relative value, which depends on its position. The figure 5 in the hundreds, he exemplified, has a relative value of 500. He said that zero was invented to fill the space of that digit whose relative value was none. For example, the number

7_3_2

(seven hundred thousand, three hundred simple units and two units) was written as

70 302.

Of course the notation 7_3_2, for the number 70302, never existed. But that is how it would be if the positional system had been created without the participation of the symbol for zero.

Finished the exposition of this group that studied the positional system, a representative of the team that wrote the work on "Zero in addition, subtraction and multiplication" was called. The illustration on the cover of the work was that of a bicycle, representing, according to the group, a pair of zeros.

The student representing this team wrote two arithmetical expressions on the blackboard and asked classmates to solve:

A) $26 - 2\sqrt{(5 + 2\sqrt{4})}$ B) $3\sqrt{[8 - 4(1 + 5:5)]}:7$

In a few minutes they found that the first expression had been zero, finishing at 26-26. The second expression resulted in zero within the bracket, and hence, it was 3.0: 7, multiplication by zero and division of zero by another number, which gives result zero.

After that, the same student wrote on the board the expression:

$3 \cdot \{4 - 5 \cdot [3 + \sqrt{2} \cdot (7-1)] + 4 \cdot (3-2)\} \cdot 0$

And asked colleagues to calculate the result, without doing

the operations on paper.

Marino noticed and immediately warned that the result would be zero, since all the expression was multiplied at the end by a zero. The value inside the braces, no matter what it can be, is multiplied by zero, and then the entire result is nullified.

Olga praised Marino for his numerical insight, commenting that it is not common for junior students to note the scope of the role of zero when written in such a position as the given expression, unless they have already been trained in that regard.

Claudia was the one who went ahead to deliver the work "Zero in the division", elaborated by the group of which Marino was part. The illustration on the front cover was the drawing of the traffic circle in front of the school building.

Claudia's exposition began by drawing the attention of the

class to the fact that the denominator of a fraction has the role of divisor in relation to the numerator. Therefore, it is forbidden to write a fraction with non-zero numerator and zero in the denominator, since it would be a meaningless fraction, corresponding to a division with zero in the key.

Then Claudia showed on the blackboard examples of possible situations for zero in the division: the division with remainder zero - exact division -, the division with null quotient and the division with null dividend. She asked the class:

- What cannot we do with zero?

- Divide another number for it - her colleagues replied.

Teacher Olga at this point drew attention saying that few people have attacked this fact. She recalled that a lot of people are afraid of doing something involving the number zero, thinking that "this cannot", that "that cannot", when in reality only what cannot be done is to divide another number by zero.

The next group to be called ahead was the one that elaborated the work with the title "Zero in the exponentiation and in the rooting". But before the student team member, Leander, started his presentation, it rang the signal for the break and teacher Olga determined that he would make the presentation the next day, when there would be the next Mathematics class.

8. Demonstration

The first information Leander gave was:

"Every number raised to zero equals one."

A class student asked:

- Any base? Even base zero?

- Even the base being zero - the representative of the team in question replied.

Olga intervened at that moment.

- This has a somewhat advanced demonstration, which requires a thorough knowledge of Mathematics. But I can make an argument to convince you.

Then Olga stood up and wrote on the board:

$x^n/x^n = 1$, para $x \neq 0$.

She asked the class:

- Does anyone disagree with what is written?

All of the students said they agreed. Then Olga wrote just below that first expression:

But $x^n/x^n = x^{n-n} = x^0$. Thus, $x^0 = 1$.

- You have seen that, at least on a non-zero basis, the statement is demonstrated — Olga continued, - assuming that the property of the division of powers of the same base is valid.

- That's the one that says... - Leander hesitated.

- That the basis should be maintained and the exponents subtracted - Olga added. Then she urged Leander to continue exposing the work.

Leander told the class that he wanted to tell them a curiosity: he said that there are only two numbers whose roots, square, cubic or larger, are equal to themselves.

- One of those numbers is "1" - Leander said. - What's the

other one?

- Zero! - The class replied in chorus.

And then Leander wrote on the board:

$\sqrt{0} = 0$.

- And how much will we have if we raise zero to an exponent other than zero? - He asked.

- Zero! - The class replied.

And he wrote on the blackboard too:

$0^n = 0$, for $n \neq 0$.

Olga took the time to explain to the group that the previous expression, $\sqrt{0} = 0$, is a particular situation of this last expression, since extracting the square root is the same as raising a number to the fraction ½. She promised that such an approach would be studied in a following grade, but she did not resist the idea of writing for the class on the board:

$\sqrt{0} = 0$ or $0^{1/2} = 0$.

9. Finishing

On Friday, the time of the last class of the week, every student in school went to the sports court. It was that there would be a football match confronting the team of the Marino's group with one of another class. Marino played center-forward, Oswald as the right-hander, Irenaeus on the far left and Evander was a goalkeeper. Julius stood in the defense.

The game was very hectic, with many threatening goals for both teams. But the goalkeepers were on their best day: no ball passed. The result was zero to zero.

All team players hugged Evander, in compliment for not having swallowed any "chicken".

- You were a monster in the goal - Marino said at the end.

- Yeah, - Evander said - now I understand the importance of zero.

@cacildo
cacildomarques@gmail.com

www.ingramcontent.com/pod-product-compliance
Lightning Source LLC
Chambersburg PA
CBHW050030230526
45470CB00003B/1216